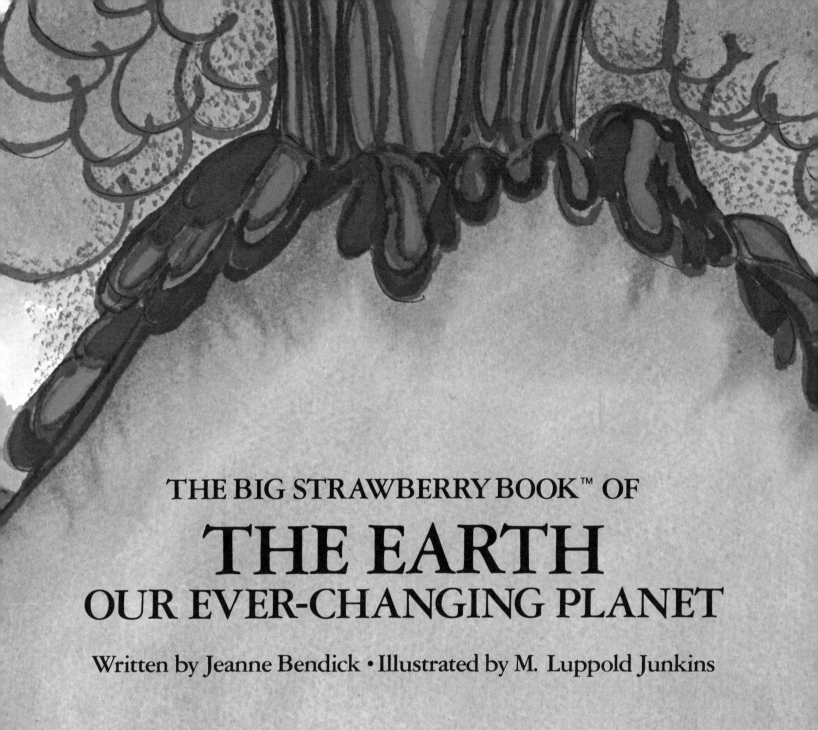

THE BIG STRAWBERRY BOOK™ OF

THE EARTH
OUR EVER-CHANGING PLANET

Written by Jeanne Bendick • Illustrated by M. Luppold Junkins

a strawberry book™
McGraw-Hill Book Company
New York   St. Louis   San Francisco   Toronto
Hamburg   Mexico

Library of Congress Cataloging in Publication Data

Bendick, Jeanne.
 The Big Strawberry Book of the Earth, our Ever-
changing Planet.

 "A strawberry book."
 1. Earth sciences—Juvenile literature.
I. Title.
QE29.B46   550   80-13908
ISBN 0-07-004514-3
*Strawberry Books™ distributed by McGraw-Hill Book Company
1221 Avenue of the Americas, New York, New York 10020*

Our Changing Earth

Contents

I. Our Changing Earth

The Earth Is Always Changing

Our planet Earth is always changing. Over the billions of years of its history, it has changed a lot.

But usually the place on Earth where you live looks just about the same today as it did yesterday. If you live where there are hills, the hills were there yesterday. But they didn't always look the way they do now. Once those hills may have been high, sharp mountains.

If you live where there are big mountains, they will still be there when you look tomorrow. But they weren't always there. Once they may have been the bottom of an ocean.

Maybe you live in the desert, but the desert wasn't always there either. Once it may have been a swamp or a jungle.

Maybe you live in the city. Once the streets were fields or forests, or maybe the bottom of a lake or an ocean.

For a long time, most people didn't believe that the Earth is always changing. They believed that the Earth had always been the same and that it always would be the same.

About 200 years ago a Scottish scientist named James Hutton watched some things that people had always watched. But he got some new ideas from watching.

He watched raindrops and saw that each

drop moved a bit of soil. He watched leaves fall and noticed that over a year or two those dead leaves became part of the soil. He watched the wind blow against rocks, and he saw where rocks had been worn away by the wind.

Hutton began to understand and to say that the Earth was always changing — drop by drop, leaf by leaf, wind puff by wind puff. And that's how it is.

Some changes are small and slow. But over millions of years small changes make big changes. Wind and water grind down mountains. Ice destroys forests. It piles up huge boulders. It digs out holes that later fill with water to become lakes. Over millions of years, rivers move mountains into the sea. Over time, new mountains are built. Continents unhitch. New oceans open. All these are slow changes.

Sometimes there are fast changes. An earthquake can make a town disappear in a few minutes. A volcano can push a new island up out of the sea or destroy an old one. A hurricane can wash away a beach or build a new one.

Many things change the Earth. Weather changes it. Forces deep in the Earth change it. Plants change the Earth. So do animals. So do people.

Clues to Long-Ago Changes

We have had to guess at many of the things we think we know about how Earth was made and about how it has changed in the past. But there have been a lot of clues. Here are some of the clues scientists have followed. Do these clues give *you* any ideas about how the Earth has changed?

All over the world, up in the mountains near the coasts, there are seashells in the soil. How did they get up there?

In Europe, North America, and Asia, even in the Sahara Desert, there are signs that once the land was buried under a thick layer of ice. How could that be?

There is coal near the South Pole. But coal began as tropical plants. How is *that* possible in a place where it is so cold that no plants grow at all?

Identical fossils of reptiles that lived 180-225 million years ago have been found in Antarctica, Africa, India, and China—places that are separated by oceans. But those reptiles couldn't swim. How did they get to all those places?

The shape of the east coast of South

10

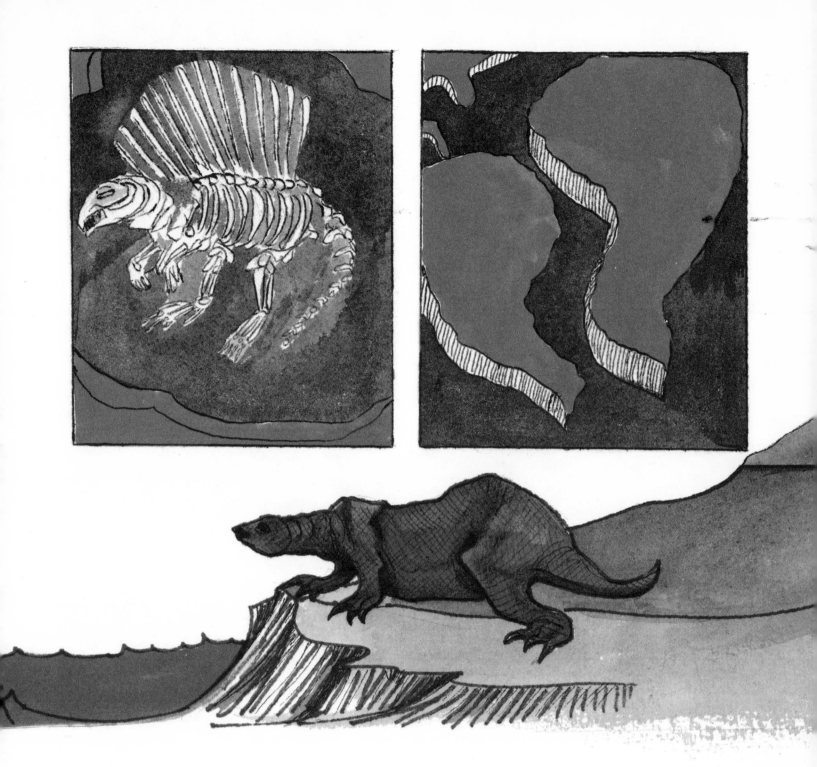

America looks as if it would fit exactly into the shape of the west coast of Africa. Is that just a coincidence?

The biggest mountain range in the world— 54,000 kilometers (40,000 miles) long — winds through all the oceans of the Earth, something like the sewing on a baseball. And the mountain chain gets bigger and wider all the time. What makes it grow?

The mountains of eastern Canada and Greenland seem to be about 4 billion years old. The rocky crust under the ocean is about 160 million years old. The Rocky Mountains, the Andes, and the Himalayas are made of still newer rock. Why are the rocks different ages?

Little by little, the Atlantic Ocean is getting wider. But the Pacific Ocean is getting smaller. What might that tell you?

Sometimes huge cracks open the Earth. Sometimes the tops of mountains blow off. Sometimes new islands pop up out of the sea. Why do those things happen? And why do they happen, over and over, in the same places?

II. The Biggest Change: Earth's Beginning

How Was Earth Made?

About 5 billion years ago there was no Earth. There was no solar system — no sun, no planets. There was only gas and dust, drifting in the blackness of space. Slowly, over half a billion years or so, the dust and gases came together into a huge cloud that flattened out and began to spin, like an immense wheel. The dust and gases at the center of the cloud came together into a giant blob with smaller blobs spinning around it. The gravities of those blobs attracted gases and solid particles from space around them.

There was so much material in the biggest blob, the one in the center of the cloud, that it all pressed together and got hotter and hotter. It started to burn and flame and glow.

That giant blob became our sun. The smaller blobs became the planets. Earth was one of them.

Before the sun caught fire, everything had been dark. Now the sun lighted up space around it for hundreds of millions of kilometers—hundreds of millions of miles. It heated the surfaces of the planets nearest it — the ones we have named Mercury, Venus, Earth, and Mars.

All the planets, from Mercury out to far Pluto, kept picking up different kinds of material from space around them. Their gravities pulled it in.

The Earth pulled in matter by chunks, pebbles, and grains. It pulled in different minerals. It attracted different kinds of

gases. Some gases combined to form water. Some water was trapped under the surface as more material piled on top.

The Earth was molten—so hot that solids melted into liquid. Slowly, over millions of years, its surface cooled and hardened. But hot material from inside kept bubbling up and piling up and spreading out to form the bumpy surface that is Earth's crust. Gases exploded up out of volcanoes and combined to make water vapor.

Thick clouds swirled around the smoking hot Earth, blotting out the sun. The vapor in the clouds came together into drops that rained on Earth, but the Earth was still so hot that the rain kept turning into steam. Then, over thousands of years, the surface of

the Earth cooled enough so the rain didn't turn into steam.

It rained for millions of years. The rain ran in rivers down the rocks and into low places to become seas. The cooling crust wrinkled and twisted into new combinations of land and mountains. But there was only bare rock and sea.

All of that took more than a billion years—about one quarter of the age of the Earth. Since then, for more than 4 billion years, the surface of Earth has changed again and again. It is still changing.

Of course nobody knows for sure if everything happened that way. Nobody was around. For most of the age of the Earth, nobody has been around.

What Is "The Earth"?

The part of Earth you see around you is called *the crust*. The crust is hills and plains, mountains and valleys. All the continents are part of the crust and so is the floor of the sea. Under a thin layer of soil or sand, the crust is all rock.

The rocky crust is called *the lithosphere*. ("Lithos" is the Greek word for stone.)

The crust looks high and deep, rough and bumpy. But if you compare the Earth to a peach, its crust is smoother and thinner than a peach skin. If you put a grain of sand on the peach, it would stick up more on the peach than the highest mountain does on Earth.

The crust is about thirty-two kilometers (about twenty miles) thick where the continents are. Under the ocean it is a little more than six kilometers (about four miles) thick.

Under the crust is a layer of the Earth called *the mantle*. The mantle makes up the biggest part of the Earth. It is like the fruit of a big peach — the part you eat.

The mantle is rock too, but it isn't stiff like the rock in the crust. It seems to be a little like clay or very thick taffy. Scientists think that the hot rock in the mantle flows, but very, very slowly. Maybe a few centimeters (a few inches) a year or less.

You can compare the layers of the Earth to the layers of a peach.

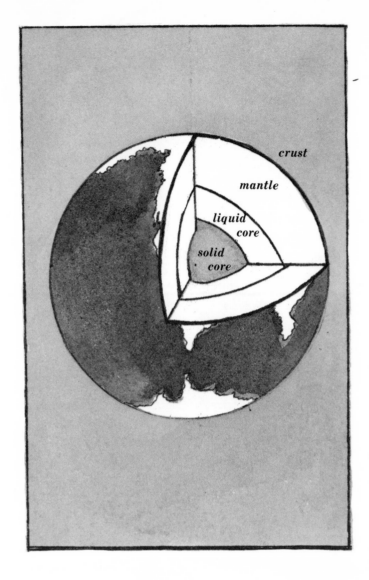

All the water on Earth — the seas, rivers and ponds, the lakes and underground water, the clouds and rain — is part of the Earth too. Together, they make up *the hydrosphere*. ("Hydro" is a Greek word for "water.")

If you washed that peach, the water that stayed on the peach skin would be much deeper, compared to the peach, than the deepest ocean is, compared to Earth. Even so, Earth is a watery planet. When you look at it from space, clouds and water are all you see. So if the water on Earth changes, Earth changes. And the water is changing.

People have explored the water and the air, but nobody knows for sure about the Earth inside. Nobody has ever been through the crust. Everything we *think* we know about the inside of our planet we have learned from the instruments that measure earthquakes and from the hot, melted rock that comes up to the Earth's surface through volcanoes.

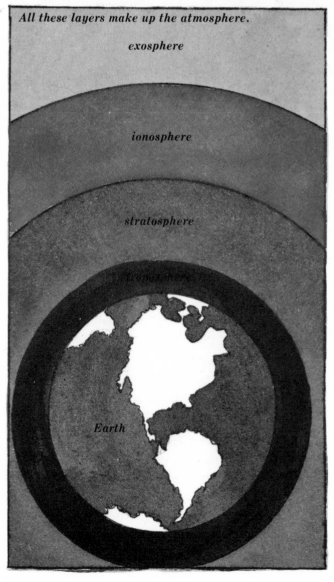

All these layers make up the atmosphere.

exosphere

ionosphere

stratosphere

troposphere

Earth

Inside the mantle is the Earth's core, which is at the center of the Earth like the pit of the peach is at the center of the peach. The outer part of the core is molten metal — flaming liquid iron and nickel. Scientists think that at its center the core is solid, squeezed in by everything above it.

The Earth is more than just what's beneath your feet. The layer of gases around the Earth is part of it too. That layer is called *the atmosphere*. ("Atmos" is the Greek word for "vapor," or "air.") Usually we call the atmosphere simply *the air*. When the Earth turns, the atmosphere turns with it.

Compared to the size of the Earth, the atmosphere is no more than a thin, delicate fuzz. From space you don't see it at all. But the atmosphere makes Earth the way it is. If the atmosphere changes, Earth will be very different. And the atmosphere is changing.

Plates: Giant Jigsaw Pieces

About fifteen years ago a Canadian scientist, J. Tuzo Wilson, was studying the places where earthquakes happened over and over. Wilson saw that earthquakes formed a pattern. They happened where large sections of the Earth's crust came together. He said that the crust wasn't one continuous, solid surface like an eggshell, but was broken into a lot of pieces that fitted together like pieces of a giant jigsaw puzzle. He called them *plates*.

Almost everybody agrees with that idea now. And we know just where most of the pieces are, where they begin and end, and

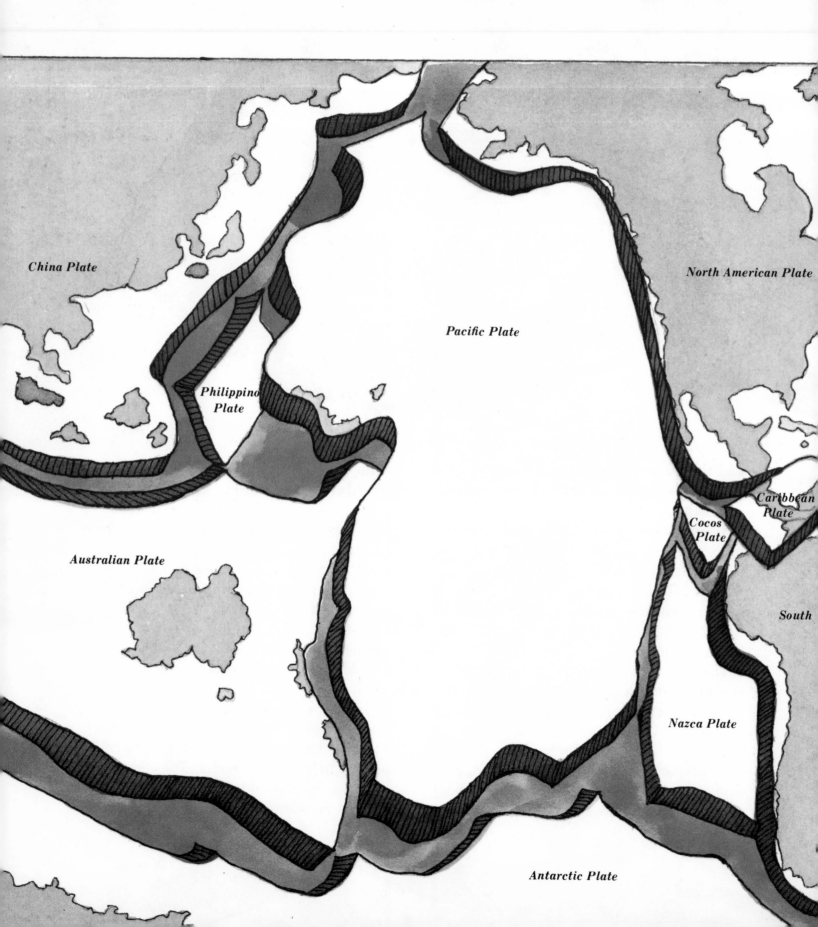

how big they are.

Some of the plates are very big. They carry whole continents or oceans. Some are smaller chips that carry a few islands or just some water.

The plates don't stay where they are, locked into their jigsaw places. They float on the mantle underneath them, moving very, very slowly as the mantle flows. But they don't all move together. Currents in the mantle move in different directions, carrying the plates with them.

At the edges of Earth's plates, there is a lot going on.

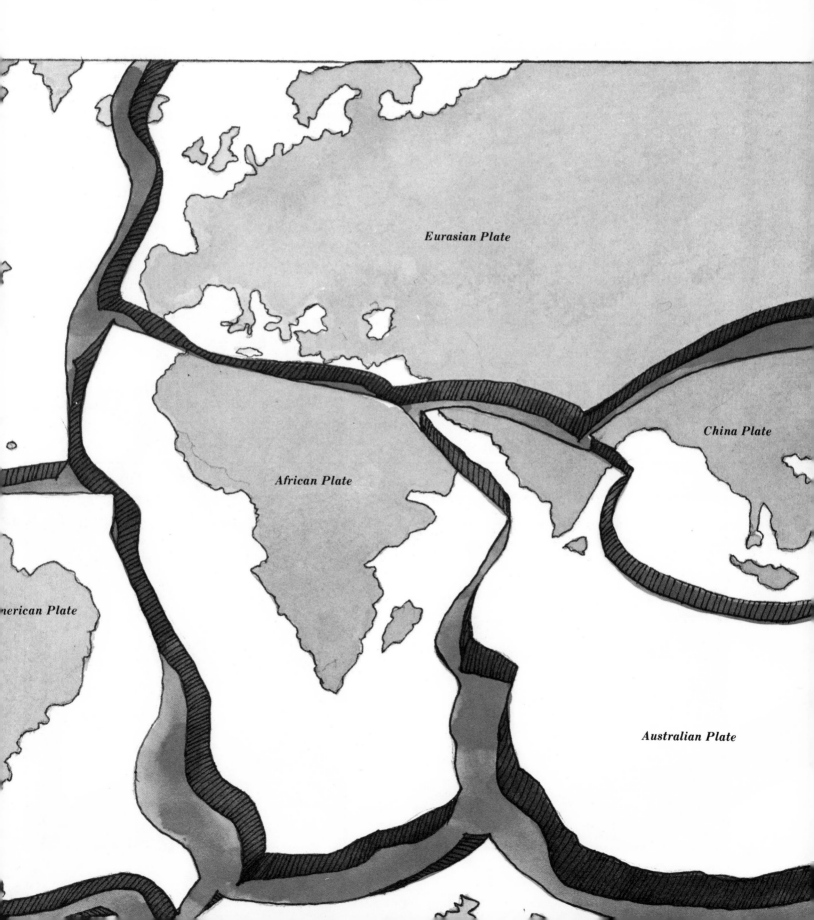

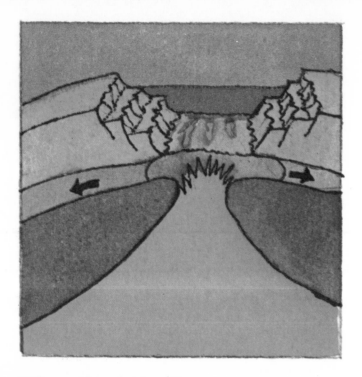

times it happens on land. The crack in the Atlantic comes ashore in Africa. As new crust forms, Africa grows. The crack comes ashore in Iceland too. Iceland grows.

Sometimes plates are moving toward each other. If this happens under the ocean, one plate dives under the other. Its edge disappears down into the mantle, where it melts. That plate is getting smaller.

After all, the Earth is a sphere with only so much surface. It can't expand like a balloon.

Where the Action Is

How the plates are moving makes different things happen at their edges.

Some plates are moving apart. When they do, new material from the mantle wells up into the crack between the plates. That's how the new mountains under the Atlantic are being built. That's how the Atlantic Ocean is getting wider.

Plates that are moving apart get bigger at their edges as the new material from the mantle hardens to form new crust. Usually that happens under the oceans. But some-

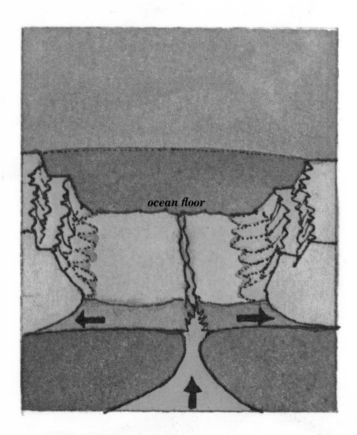

ocean floor

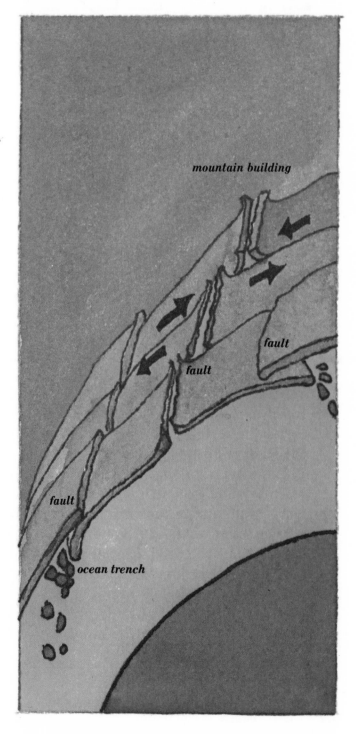

kinds of Earth rock. So when plates carrying continents collide, the edges are slowly pushed up into new mountains.

Sometimes plates are moving past each other in different directions. Where they come together is called *a fault*. The San Andreas fault in California is where the Pacific Plate comes ashore. The Pacific Plate is creeping north. But the American Plate, next to it, is slipping west. What happens? EARTHQUAKES!

mountain building

fault

fault

fault

ocean trench

Its crust has to fit it. So if the crust gets bigger in one place, it has to get smaller somewhere else.

A diving plate makes a very deep place in the ocean. Those deep places are called *ocean trenches*. Wherever there's a trench, you know that one plate is diving under another.

A plate that is carrying ocean can sink under another plate because oceans and the rocky crust of the ocean floor are heavy. But if a plate is carrying a continent, it can't sink. Continents are light, compared to other

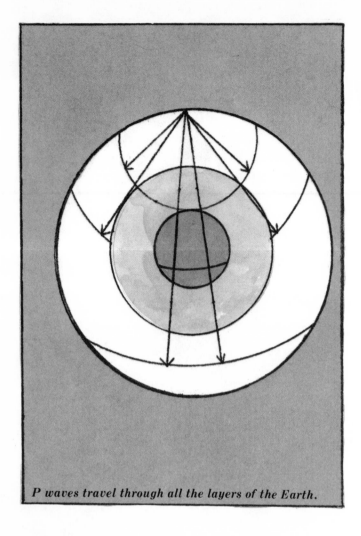

P waves travel through all the layers of the Earth.

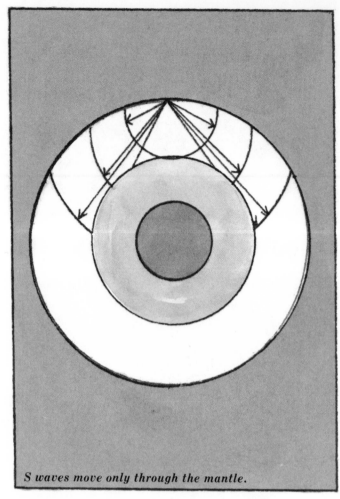

S waves move only through the mantle.

Earthquakes Change the Earth

Earthquakes and volcanoes are huge releases of energy from under the Earth's surface. Most earthquakes and volcanoes happen around the edge of the Pacific Plate — the plate that carries the Pacific Ocean. The rim of the Pacific Plate is called *the Ring of Fire*.

Another big earthquake zone stretches from southern Europe and northern Africa across the Middle East to Asia where it joins the Ring of Fire. It's called *the Alptide Belt.* Plates meet all along the Alptide Belt too.

An earthquake's energy takes the form of waves that vibrate back and forth. These vibrations start when rocks are either squeezed too much or stretched too much along a fault. Where the rocks break under the surface is called the earthquake's *focus.* The place on the Earth's surface right above the focus is called *the epicenter.* That's where the biggest damage occurs.

Other earthquakes begin far below the crust — down in the mantle. Nobody knows what causes those.

Most of what we know about the inside of the Earth we have learned from instruments called *seismographs.* These instruments track and measure earthquake waves. (Earthquake waves are called *seismic waves.*) There are three kinds of earthquake waves.

Primary waves, or P waves, start when the earthquake starts. They move away from it, traveling fast — about eight kilometers (five miles) a second. P waves move easily through all the layers of the Earth. They push straight through solids. They bend when they push through liquids.

Secondary waves, or S waves, travel only half as fast as P waves. They don't pass through liquids so they can't move through the outer core. They shake the rocks in the

22

L waves travel along the surface of the Earth.

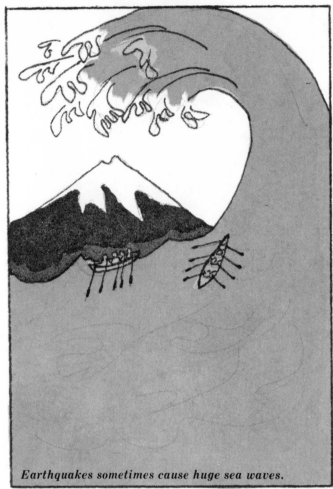

Earthquakes sometimes cause huge sea waves.

mantle and the crust sideways.

Surface waves, or L waves, travel along the surface of the Earth. The stronger the L waves are the more damage the earthquake does. L waves split the ground and shake buildings apart and twist up roads and bridges. L waves are slow. They are the last to reach the seismograph.

The speeds of the different kinds of waves and the ways they move, bend, or disappear are registered on seismographs all over the world. The seismograph pictures of those waves have drawn a map of the inside of the Earth from the crust to the core. That's how scientists have figured out what it's like in there.

Earthquakes help to build mountains by pushing the crust up into folds or blocks. Over millions of years, the mountains get higher.

Earthquakes that begin on the ocean floor sometimes cause huge sea waves called *tsunamis*. Sometimes tsunamis are called *tidal waves*, but they have nothing to do with the tide.

Tsunamis travel away from the earthquake, sometimes as fast as 800 kilometers (500 miles) per hour. They can be as high or higher than a six story building.

As a tsunami comes close to shore, the sea is sucked away from the coast, then rushes back in a series of giant waves. In 1755 the city of Lisbon, in Portugal, was destroyed in six minutes by an earthquake that was followed by a tsunami. Sixty thousand people were killed.

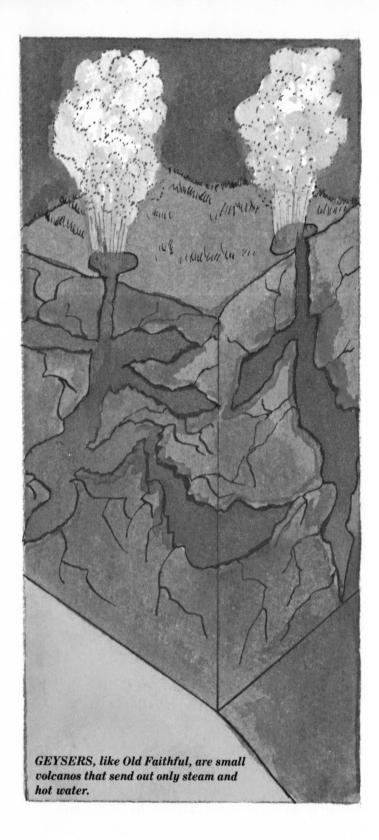

GEYSERS, *like Old Faithful, are small volcanos that send out only steam and hot water.*

Volcanoes Change the Earth

Earthquakes happen where plates are pushing together; where plates are moving apart, the spaces they open up are called *volcanoes*. A volcano brings up material from the mantle — molten rock called *magma*.

A large part of the Earth's crust was formed by volcanoes. Sometimes, where plates are drifting apart, the magma just keeps bubbling up and cooling to make new crust. But in some places where the magma

can't flow out slowly and cool, it collects until it explodes.

When a volcano under the sea slowly bubbles and cools, the magma piles up into undersea mountains. When those mountains get tall enough to stick out of the water, they become islands. The Japanese islands are volcanic islands. So are the Aleutians.

The Hawaiian Islands are volcanic too, but scientists think they are being built in another way. They think that under the Pacific Plate there are permanent hot spots

SHIELD VOLCANOS *are broad, gentle hills, formed by flowing lava.*

CINDER CONES are steep, cone-shaped mountains like Mt. Fuji and Mt. Vesuvius.

other gases, ashes, stones, and lava. All of that is smoking hot, but a volcano does not throw up fire. Sometimes material from inside the volcano is thrown fifty or sixty kilometers (thirty or forty miles) into the air. Sometimes it makes the day as dark as night. Sometimes the dust and ashes from a volcano blow around the world.

In 79 A.D. the volcano in Mt. Vesuvius, in Italy, blew up and destroyed the city of Pompeii.

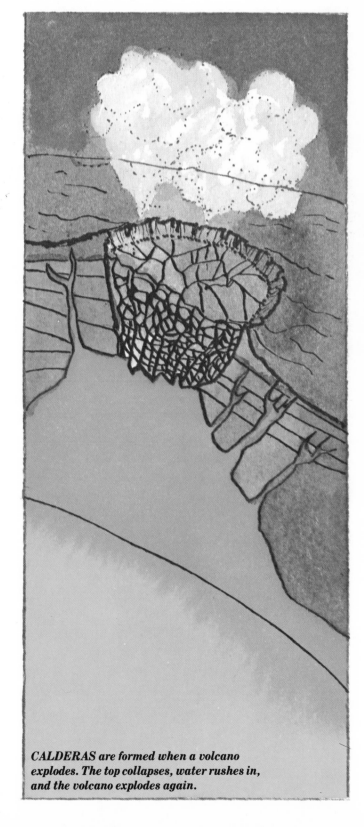

CALDERAS are formed when a volcano explodes. The top collapses, water rushes in, and the volcano explodes again.

in the mantle that push up volcanoes in the plate drifting over it. There seem to be a number of hot spots in the mantle. Nobody knows what causes them.

The new island of Surtsey, on the north end of the Mid-Atlantic Ridge, is an exploding volcano. One way volcanoes change the Earth is by building new crust out of material from underneath the Earth's surface.

Another way they change it is by blowing up mountains and islands.

A volcano on land throws up steam and

In 1883 the volcano on the island of Krakatoa, in Indonesia, blew up. The sound was so loud it traveled around the Earth six times. Dust and ashes drifted through the atmosphere for years. (That made sunsets very beautiful.) But the tsunami the volcano caused drowned more than 30,000 people.

The biggest volcanic explosion we know about makes Krakatoa seem like a very big firecracker.

In 1500 B.C. the volcano on the Greek island of Thera, in the Mediterranean Sea, blew up and destroyed one of the great civilizations of the ancient world — the Minoan Empire. Islands for miles around were covered with hot ashes. Cities and palaces were destroyed by the huge wave that traveled out from the volcano. All the ships in all the harbors were sunk. There was no more Minoan Empire.

There are about 500 active volcanoes on Earth. "Active" means that something is still going on in them. They are still bringing up magma, or there is a chance that they may explode at some time.

Ever since the Earth began, there have been earthquakes and volcanoes. They change the Earth in big ways. People have no control over them, at least not yet. We can't control changes that begin below the crust. They are too big.

IV. Changes That Begin Above Earth's Crust

Weather and Climate Change the Earth

Some changes begin above the Earth's crust, in the atmosphere. That's where the weather begins.

A lot of things make up the weather—how much the sun shines; how hot the land, air, and water get; how hard the wind blows and from what direction; how much it rains.

Those things are all connected. Which way the wind blows decides whether it will rain or not, and *that* decides if a place will be jungle, grassy plains, or a desert. The weather changes the Earth.

The weather changes with the seasons and the seasons change because of the way the Earth is tilted toward the sun or away from it as the Earth orbits the sun every year.

Weather and climate are not the same. Weather is how hot it is, how much it rains, how windy it is *at a particular place and time*. Climate is the kind of weather the whole Earth, or a particular place on Earth, has *over a long time*.

28

Every place on Earth is different because its climate is different. The climate of a place can cover a whole country or even a continent. But smaller places have their own little climates.

A big field doesn't have the same climate as the forest next door. The climate at the top of a mountain is not the same as the climate at the bottom.

There are many climates around today's Earth, but that wasn't always so. Once a great part of the Earth was a humid swamp.

At other times most of the continents were deserts. And for long periods, most of the Northern Hemisphere was covered with ice.

Those changing climates changed the Earth in many ways. As so much of the sea froze, then melted, the shorelines of the continents changed.

As the climate changed, different kinds of plants grew or died. So did different kinds of animals. What made the climate change so much? Nobody is sure. But one reason may have been changes in the wind.

29

wind

soil

The wind blows soil around the Earth.

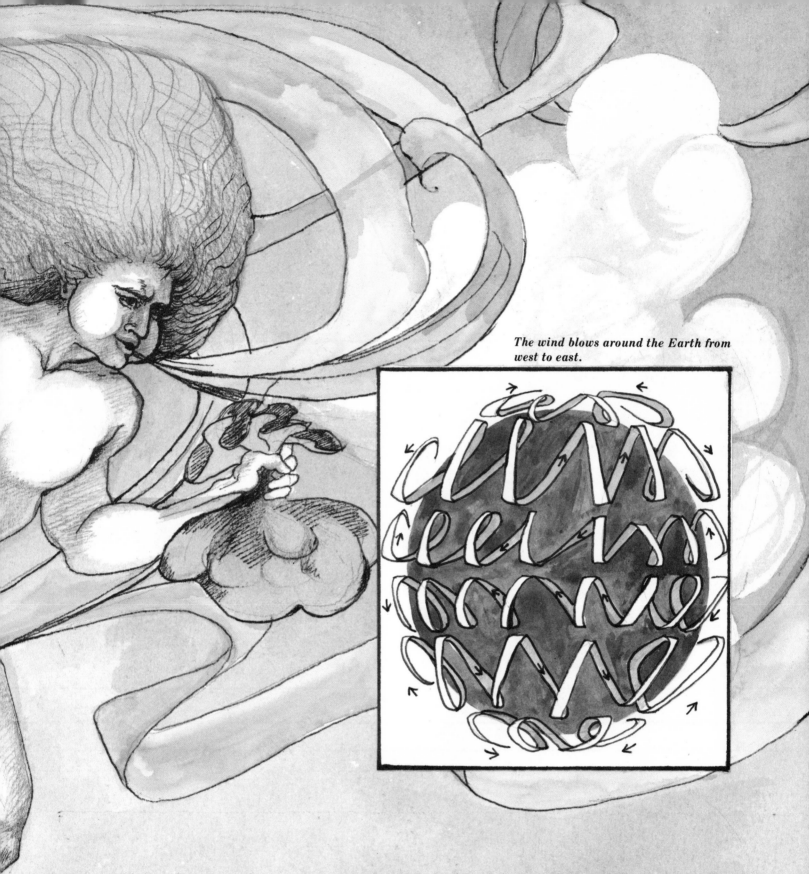

The wind blows around the Earth from west to east.

Wind Changes the Earth

If there were no air on Earth, there wouldn't be any wind. Wind is moving air. The spinning Earth and uneven heating by the sun make the air move. And the wind changes many things.

The wind blasts rocks with sand and grinds them up to make soil. Then the wind carries that ground-up rock to other places and dumps it. Wind-deposited soil is called *eolian soil* after Aeolus, the Greek god who was supposed to rule the wind.

The wind blows from west to east around the spinning Earth. When water evaporates into the air, the wind moves that water around the Earth until it falls as rain or snow. Where it falls makes a place a desert, a jungle, or farmland.

How hard the wind is blowing, and where, can make a hurricane or a blizzard out of an ordinary storm.

31

The same water keeps going up and coming down.

The Water Cycle

All the water that was ever on Earth is still here. It cannot get away. But we can't make any new water, either. And there's no way to get any from outside our atmosphere.

There seems to be plenty of water in the oceans, in lakes and rivers, and stored underground. More is locked up in ice at the North and South Poles. Some is locked into the rocks, where we can't get at it.

Most of the water on Earth is salty. Land plants, animals, and people can't use the water unless it is fresh. They can't use it if it is polluted, either.

Wonderfully, the water on Earth moves in cycles. The sun pulls water vapor up into the air from the ground and from plants and animals. It pulls water up from puddles and rivers, lakes and oceans. Most of the salt is left behind. So is most of the pollution.

The water vapor condenses into clouds and the wind moves the clouds to other places. When the water falls again it is fresh and clean. The water cycle takes care of that.

People are beginning to change Earth's water so much with different kinds of pollu-

tion that the water cycle can't clean it as fast as people make it dirty.

Irrigation water full of chemical pesticides and fertilizers flows into the rivers and out into the oceans. Factories pour chemicals into rivers and lakes. Oil spills. Garbage gets dumped.

The water cycle can't clean all that pollution out. In some places the rain has acid and other chemicals and radioactive materials in it.

People have changed the water. What will happen to the Earth if the water cycle stops working? Could plants and animals adapt to living with polluted water? Could people unpollute the water with chemicals?

Or do you think controlling pollution is the best idea? In many places people are trying to fix their past mistakes. Lakes and rivers that were poisoned and "dead" are beginning to live again.

How a River Changes the Earth

The Colorado River carved the Grand Canyon more than one and a half kilometers (about a mile) deep. It took 10 million years, but that's a short time in the age of the Earth. A river digs a ditch for itself all the way from its beginning to its end where it flows into the sea. Rivers do that all over the world.

Usually rivers start as springs or tiny brooks up in the mountains. Brooks join to make streams. A fast-moving stream carries sand, pebbles, stones, and even big rocks. Rocks bump against the banks of the stream and wash away soil.

Streams join to make rivers. Rivers join to make bigger and bigger rivers. And all along the way they pick up more soil and stones as they flow downhill to the sea.

All the solid material that a river picks up and carries is called *sediment*. If the sediment catches on rocks and roots, it can pile up to make an island in the river.

Where a river flows into the sea is called its *mouth*. The mouth of a slow-moving river is usually a big stretch of calm water. The river drops its sediment there. In quiet water, all that mud and sand pile up to form new land. That's called *a delta*. The Mississippi River has built a huge delta where it flows into the Gulf of Mexico. The River Nile, in Egypt, has built a delta too.

Where a fast-moving river empties into the sea, the water is deep because the river car-ries its sediment out into the ocean. The river mouth stays wide and deep. The world's great seaports are often at the mouths of rivers like that.

But the sediment that's dumped into the sea doesn't go away. It is dumped into deep places, usually where undersea plates come together. Slowly, the sediment is squeezed up into ridges that finally become mountains. Some day that sediment will be dry land again.

What Makes a Desert?

There's a lot of desert on Earth. About one third of all the land is either desert or almost desert. And the deserts are growing.

Any place on Earth is considered a desert if it gets less than twenty-six centimeters (about ten inches) of rain a year. Some deserts get less than a centimeter (about half an inch) of rain a year. Some go for years and years with no rain.

Some deserts are all sand—oceans of sand where nothing grows at all. Some deserts are nothing but bare rock. The wind keeps grinding the rock to sand, then blowing the sand away. Some deserts have a few plants growing—cacti and thorny bushes.

Not all deserts are hot. Some are very hot during the day and very cold at night. Some are near the North and South Poles, where it is always very cold.

Any place on Earth can become a desert if no rain clouds form over it and no rain falls. One thing that keeps rain away is high air pressure.

Air pressure is the force with which the air presses on a surface. Since air is made of particles, the more air particles there are, the heavier the air is and the more it presses. That heavy air stays close to the ground. The air pressure is high.

When there are fewer particles, the air pressure is low. The air is light, so it rises

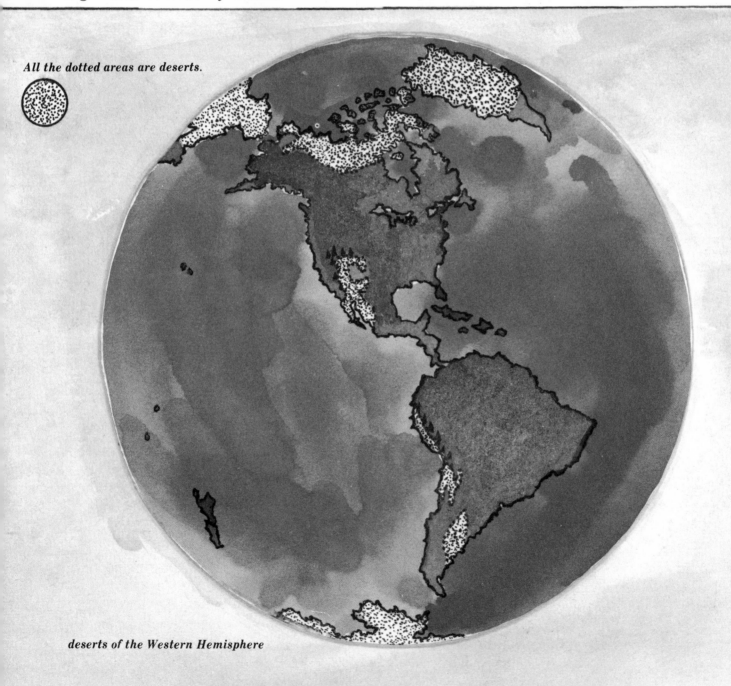

All the dotted areas are deserts.

deserts of the Western Hemisphere

away from the ground. As it rises, it cools. When air is cooled, the water vapor in the air condenses into clouds that bring rain. So rain clouds form when the air pressure is low, but they don't form when it is high.

Around the equator the air pressure is low and it rains a lot. That's where the tropical jungles are. But above and below that zone, the air pressure is high so it doesn't rain. That's where the biggest deserts are, in belts around the Earth.

Sometimes mountain ranges can make deserts. If the winds that carry rain have to climb up and over mountains, the air gets cooler and cooler as it climbs. The water vapor condenses into raindrops, and it rains.

All the rain falls on one side of the mountains and never gets over them to the other side. So the other side is a desert.

Some of the driest deserts in the world are right at the edge of the ocean, because cold ocean currents can make deserts too. Very cold currents of water flow from the South Pole up the western sides of South America, Africa, and Australia. Ocean winds, carrying rain toward the continents, blow across those currents. But the currents cool the winds, and they drop their rain before they get to the shore. The land along the coast becomes a desert.

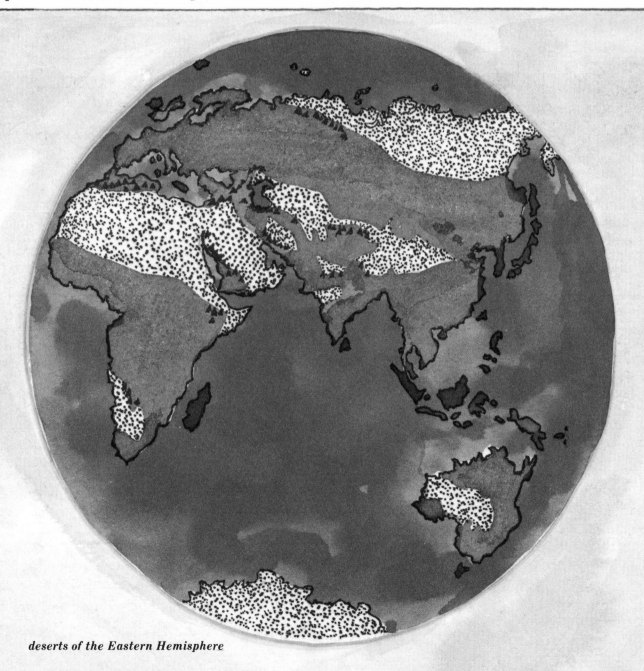

deserts of the Eastern Hemisphere

People make deserts too.

In dry places, where it's hard for plants to grow anyway, sometimes people own too many plant-eating cattle. The herds eat everything — grasses, scrubby bushes, even thorny trees — until there are no plants left.

Once the deserts of Texas were covered with grass. Then the settlers let too many cattle graze there.

Once parts of the African desert had grass and trees. But too many cattle ate the grass. People cut down the trees for firewood. Now the African desert, the biggest in the world, is growing bigger by about forty-eight kilometers (thirty miles) a year.

Bad farming can change fertile land into a desert.

When the American Great Plains were plowed up and dug up to plant miles of wheat, a long dry spell in the 1930s blew away the topsoil and the land was a dusty desert for years.

Planting the same crops over and over in the same soil can make a desert. The nutri-

40

ents in the soil get used up. The soil can't support plants. When no plants grow, the soil washes away or blows away.

In some places the wrong kind of irrigation raises the level of salt in the soil so that nothing will grow there. That's happened in parts of Africa, parts of Pakistan, and parts of Mexico. And the deserts grow.

Sometimes chemicals—pesticides, fertilizers, chemical wastes, or radioactive wastes—poison the soil. They kill the tiny organisms that make soil good. Without those or-

ganisms plants won't grow. So the desert does.

Once a large amount of land becomes a desert, there's almost no way to keep the desert from growing. If plants aren't there to hold the soil, the soil dries into sand and the wind blows it away. The blowing sand covers more plants and grinds more rocks into sand and the desert grows.

Some scientists think that someday deserts will cover most of the Earth.

Once this much of the Earth was covered with ice.

Ice Changes the Earth

Even a little ice can change things. Ice on branches can topple trees. Water freezing in the soil lifts it up. Water freezing in rock cracks can split the rocks.

Sometimes in the history of Earth there has been a *lot* of ice. Five times in the last 2 million years or so huge sheets of ice have spread out from the north to cover one quarter of the Earth. Nobody is really sure why that happened. Some scientists are beginning to think that regular, slight changes in the Earth's orbit caused all the ice to form.

Some scientists think that the rising and falling of great mountain chains changed the pattern of winds blowing around the world, and that that changed the Earth's climate. Some think that the drifting continents changed the flow of the great ocean currents.

Some think that changes in the atmosphere made the Ice Ages. Maybe great volcanic eruptions blew so much dust into the atmosphere that the dust blocked out the sun's heat and light.

Others think that the causes were far away from Earth itself. Maybe there were changes in the amount of radiation from the sun. Or maybe the whole solar system passed through immense dust clouds in the galaxy and those clouds blocked off the sun's heat.

In the first Ice Age, sheets of ice thousands of meters thick covered one third of the land area of Earth for 100,000 years. So much of the water on Earth froze that the level of the oceans fell. Land that had been under water for millions of years became dry.

As the ice caps ground across the continents, they scraped down mountains and pushed soil and rocks ahead of them. The ice dug holes in bare rock that later filled with water to become lakes. The ice ground up rock into soil, which the wind carried around the world.

Five times the ice formed; it melted when the climate of the Earth got warmer, and then reformed when it got colder.

We are now living at the end of the fifth Ice Age. The ice has been melting back toward the poles for 10,000 years. Will another Ice Age come? Probably it will. But you don't have to worry that you'll be in it. Usually, it is about 150 million years between Ice Ages.

Ice is still changing the Earth today. In the most northern parts of the world it never

melts. Slow-moving rivers of ice called *glaciers* creep down mountains. They grind rocks and dig cracks and cliffs. They make new valleys.

The ice cap at the South Pole is much bigger than that at the North Pole. It's twice the size of Australia, covering the continent of Antarctica. Underneath the ice there are mountains we have never seen. If the ice at the poles ever melted, the coasts of all the continents would be underwater. If the ice cap grows, the water level would fall all over the world and today's seashores would be miles away from the sea.

Could people cause the polar ice caps to grow or to melt? Scientists think they might, if they change the Earth's air too much. And the air is changing.

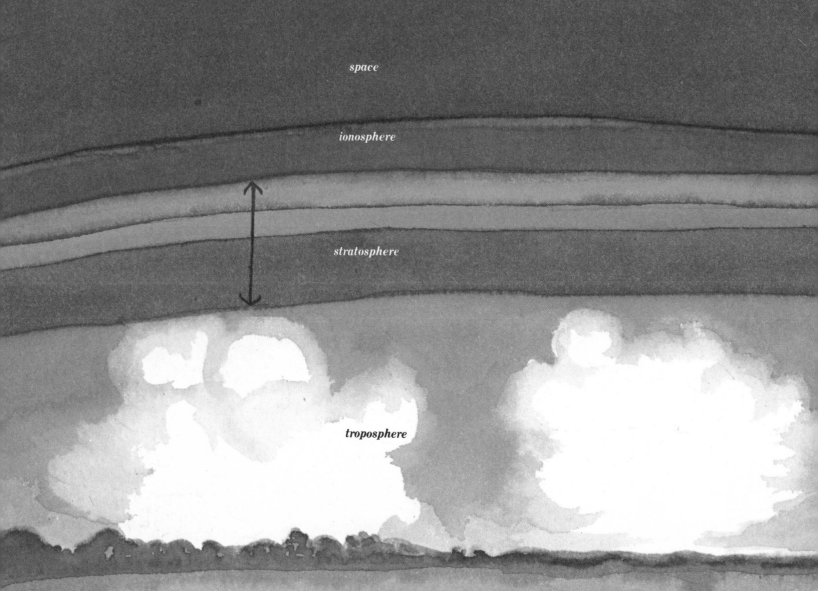

space

ionosphere

stratosphere

troposphere

The Air Blanket

Earth's atmosphere is like a thin blanket around it, but what a magic blanket! It protects Earth from the sun. Without that air, the sun's radiations would be deadly to all life on Earth.

The atmosphere keeps the Earth from getting too hot or too cold.

It supplies the gases that all living things depend on.

The atmosphere isn't the same from top to bottom. It is heaviest from the Earth's surface to about sixteen kilometers (ten miles) up. That layer of air is called *the troposphere*. All living things live in the troposphere. Most of them live in the bottom five kilometers (about three miles). Above that there is not enough oxygen to keep most animals — including people — alive.

Land animals and sea mammals absolutely depend on the oxygen in the air. Animals that live underwater absolutely depend on the oxygen in the water. Animals can't live for more than a few minutes without oxygen.

But the oxygen wasn't always there. It wasn't part of the atmosphere until the plants appeared on Earth. Just about all the oxygen in the atmosphere is made by plants.

On land, it is made by trees and grass and weeds and cacti and all other plants. In the oceans it is made by algae, from the tiniest, one-celled plants to the biggest kelp.

Every green plant, from the biggest tree in the forest to the tiniest green speck in the sea, puts out oxygen while it is making food for itself.

The air around us is made of a mixture of gases. It is about 78% nitrogen, about 21% oxygen, and the rest is tiny amounts of other gases. There are also other things in the air. There is water vapor, some salt from the

44

oceans, dust and gas from volcanoes, and pollen and spores from plants. These things have been in the atmosphere for millions of years, and they haven't really changed it.

But people have made a fast change in the atmosphere, especially in the last 200 years since they have been burning coal and oil. Now the air is full of smoke, gases, and chemicals from cars, factories, chimneys, and power plants. It's full of insecticides and aerosols.

Most of those things are harmful for people and other animals to breathe. Some of them kill plants. Some ruin buildings. Some put holes into the upper layers of the atmosphere that protect the Earth from the sun.

Over some cities the air pollution never goes away. From a plane you can see it for miles, a thick brown or yellow cloud hanging over the city. Once the sky there was clear. Now it has changed.

People have put radioactive material into the air too. Some of it is taken in by people and other animals. Some is taken in by plants. Some of it falls with the rain.

Nobody is sure yet whether the changing atmosphere is going to make big, fast changes on Earth. Some scientists think it may upset the whole balance of heat on our planet and so change the climate everywhere.

Maybe pollution will keep enough sunlight from reaching Earth. Then the climate will get colder. Or maybe the layers of polluted air will let heat in and keep it in, the way glass does in a greenhouse. Then the climate would get hotter and hotter and more polluted.

People have changed the air. Nobody is sure yet what effect that will have or if it can be changed back to clean air again.

V. All Living Things Change the Earth

Neighborhoods Depend on Each Other

Our environment is where we live. It is everything — air and water, soil and rocks, sunlight, wind, weather, buildings, machines, and all living things. The environment of Earth is a very big neighborhood. But neighborhoods can be small too.

A neighborhood can be a city block, a park, or the woods. It can be a backyard. The ocean is one kind of neighborhood. A beach is

another. The different kinds of living things in any neighborhood depend on each other, but no neighborhood can get along by itself. It needs things from outside.

Smaller neighborhoods depend on the bigger ones around them for some of the things they need. The tiny plants and animals in a drop of pond water depend on the bigger neighborhood of the whole pond.

Caterpillars, aphids, bark insects, and birds can live together in a tree neighborhood. But the tree depends on the sun and the rain and the soil of the bigger neighborhood around it for the things *it* needs.

Scientists call the living things in each of those small neighborhoods *a community*. A number of communities living side by side and depending on each other are called *an ecosystem*.

Neighborhoods Change

It's natural for neighborhoods to change. They always have. Usually they change slowly.

So many plants might grow in a pond that the pond fills in. Ferns grow where the pondweeds grew. Then small trees begin to grow. Then different kinds of larger trees.

What was once a pond will become one kind of forest, then another.

Then maybe the trees will die because beavers make a new pond there, so the cycle of plants in that neighborhood starts again.

Other kinds of neighborhoods change too. Beach neighborhoods change into grassy dunes, or maybe into wetlands. Prairies

change into deserts. And every time the plants in a neighborhood change, the animals change, because the animals depend on the plants. They always have.

People change neighborhoods. Often they change them fast. Sometimes they make plants grow where they have never grown before. But usually they clear away the plants and change a plant neighborhood into a city neighborhood.

A city neighborhood isn't different from any other kind of neighborhood. It depends on the bigger neighborhoods outside itself for the important things it needs — air, water, sun, and food and fuel that began with plants.

Plants Change the Earth

Nobody knows for sure how living things started or just what the first ones were like. Scientists think they were tiny, one-celled specks floating in the warm seas of Earth, billions of years ago.

One kind of speck became able to do something that changed the Earth forever. It found a way to make food for itself, in itself. Those tiny food factories were the beginning of plants.

The smallest green plant that floats in the sea, even if it is smaller than a pinpoint, can make food for itself. No animal can do that.

When an animal needs food it has to eat plants or other animals.

Green plants make food for themselves out of sunlight, water and minerals, and carbon dioxide and other gases from the air. When plants make food for themselves they give off the oxygen that animals need to live.

At first, all the plants were in the water. They were simple, one-celled algae. After a while — millions of years — some cells joined together to make bigger, more complicated plants. Plants spread from the sea into shallow bays and then into swampy places.

But on land, there was still nothing but bare rock and stones and pebbles and sand.

52

Nothing grew. Nothing was alive on land.

Then, little by little, the plants moved up out of the edges of the shallow water and the swamps onto the shore. They were small, simple plants that had to grow on bare rock because there was no soil.

Those first plants —algae, lichens, mosses —looked like smears of color on the bare rock. There are still plants like that around.

When those plants died, and others grew on top of them, and on top of them, the soil began. Soil is made partly from living things that have died and partly from rock that has been crumbled by water and weather.

Then there were ferns and bigger ferns, and after thousands of years there were great forests of ferns and huge plants that looked like trees but weren't like the trees we have now.

For hundreds and thousands of years those huge plants grew and died and fell and rotted and other plants grew and died and fell on top of them. It all pressed down and down until it became a rich, compacted vegetable material called *peat*.

Whole forests in swampy places sank and were buried. They sank deeper and deeper. Layers of sandstone and shale rock formed over them, squashing those ancient forests into coal.

Nobody is quite sure how oil and natural gas were formed, but most scientists think they began with billions of tiny plants and animals that lived in ancient oceans. As they died and settled to the bottom, bacteria caused their remains to decay. Sediment piled up over them for millions of years. The sediment was pressed into rock which pushed down even harder. The pressure of rock, heat, and the action of bacteria changed those tiny ancient living things into oil and gas. Over a very long time the oil and gas were sealed into pools, under the ground and under the oceans.

In the first great forests, everything was green. Green leaves and green stems. No brown or gray trunks. No flowers. No seeds. No fruit.

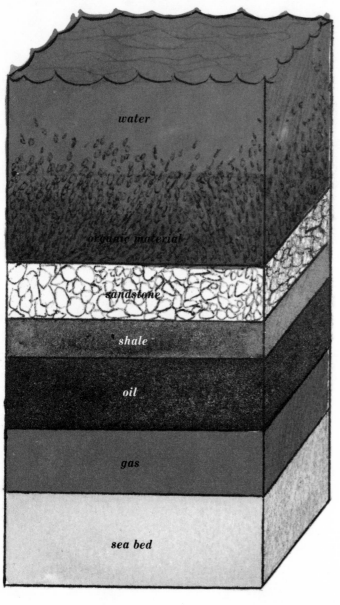

water

organic material

sandstone

shale

oil

gas

sea bed

Then there were real trees, something like pines. They had cones. The cones had seeds. Seeds were the beginning of the kinds of plants we have now. There are thousands of kinds of plants, and they have changed the Earth in many ways.

Without the plants there would be no soil. There wouldn't be any oxygen in the air for animals to breathe, but there wouldn't be any animals either because animals depend on plants for food. Plants are the most important producers on Earth. They make the food and oxygen. Animals are the consumers. They use what the plants make.

Plants keep changing. Mostly they change naturally.

early tree

They change when the climate changes. Plants that need a lot of rain can't grow if the climate changes from wet to dry. Plants that grow in warm places can't live if the climate gets colder.

When plants change the whole neighborhood changes because animals live where their food is. Animals eat plants or they eat animals that eat plants. But different kinds of animals eat only certain plants. So when plants change the animal neighbors change too.

Animals Change the Earth

In every ecosystem, every kind of living thing does something useful for the neighborhood. Plants do. So do animals.

The most useful animals are not the biggest or the most important looking.

Earthworms turn the soil over and make it richer with their juices. Ants and other underground animals make air spaces in the soil. When they die they make the soil richer.

Under fallen trees and rotting leaves are all kinds of creepy-crawly things that help that dead material decay. If dead plants and animals didn't decay they would keep piling up until there was no room on Earth for living things.

Other kinds of animals — vultures and crows, snails and hyenas — clean up dead animals and other wastes.

Bees and flies may be pests to people but some plants can't make seeds without them. Many animals carry seeds. This helps plants get planted in new places.

Sometimes animals change the neighborhood. Coral animals build islands and seashores. Beavers build ponds and change the direction of streams. Some insects can destroy an orchard or a woods. But most animals don't change their neighborhoods much.

Except one.

People Change the Earth

People haven't been around for long, as Earth history goes. But they have made very big changes. They have changed the environment faster than earthquakes, volcanoes, storms, droughts, or Ice Ages.

People have reshaped the environment for themselves. They have cleared millions of square miles of land, first to raise crops, then to build towns and cities. They have paved over the land with roads to everywhere.

They have chopped down thousands of

miles of forests to get wood for heating and building and making paper.

People have dug up the Earth for metals and drilled down into the crust for coal and oil. In some places they have dumped radioactive waste, which has become part of the chemicals in the soil.

People have changed the course of rivers to irrigate their crops and stopped up rivers with dams to make electricity.

People have changed the soil, the water, and the air. Those are big changes.

Slow Changes, Fast Changes

If things are left alone, most of the changes on Earth are slow. They take hundreds or thousands or even millions of years. The Earth changes. The shapes of land and oceans change. The climate changes and that makes plants and animals change. It's natural for things on Earth to change.

Those changes have always happened slowly. They have happened so slowly that there was time for living things to become adapted to the changes.

But people have changed the environment of Earth fast. They've cut down forests and mountains. They've moved rivers and shore-lines. They've changed the air and the water.

When people make quick changes in the environment of Earth they push and pull on the web of life. A web is delicate. Every part of it is attached to every other part. If one

strand breaks, sometimes the whole web falls apart.

When people make quick changes on Earth, sometimes they don't think about the consequences. Sometimes they do think about them but they guess wrong about what the consequences will be. They don't know what is going to happen until after it has happened. Then, sometimes it's too late. You can't always go back and undo a change.

Some people think that if humans aren't more careful, Earth might become like the other planets in our solar system — a desert place with poisonous air and water. And no life. But some people think that maybe we are beginning to understand the web of life on Earth. That it's not too late to turn things around and change Earth back to the way it was.

59

Can People Change Other Planets?

Earth is a very special planet. It has fresh air and clean, liquid water. It has sunlight and life. No other planets or moons in our solar system have those things. If they have an atmosphere it is made of poisonous gases. If there is water, it is frozen. If it rains, the rain is acid.

60

Someday maybe people will be able to change those moons and planets. Maybe we will be able to make things grow there. Then there might be air to breathe and water to support life. Maybe humans and other animals could live there. Maybe Mars or the moon could become a neighborhood. Like Earth.

VI. Index